C0-GAH-886

Jane Goodall

Living Chimp Style

by Mary Virginia Fox

Illustrated by Nona Hengen

DILLON PRESS, INC. MINNEAPOLIS, MINNESOTA

Library of Congress Cataloging in Publication Data

Fox, Mary Virginia.
Jane Goodall, living chimp style.

(Taking part; 10)
SUMMARY: A biography of the first scientist to study chimpanzees in their natural habitat.
1. Lawick-Goodall, Jane, Barones van—Juvenile Literature.
2. Chimpanzees—Behavior—Juvenile Literature.
3. Zoologists—England—Biography-Juvenile Literature.
4. Mammals—Behavior—Juvenile Literature.[1. Lawick-Goodall, Jane, Barones van. 2. Zoologists. 3. Chimpanzees] I. Hengen, Nona L. II. Title.
QL31.L38F69 599.88'44'0924 [B] [92] 80-27542

ISBN 0-87518-204-6

©1981 by Dillon Press, Inc. All rights reserved
Second printing 1982

Dillon Press, Inc., 500 South Third Street
Minneapolis, Minnesota 55415

Printed in the United States of America

JANE GOODALL

Dr. Jane Goodall is known worldwide for her contributions to ethology, a relatively new science which is the study of animal behavior. For the past twenty years she has observed the chimpanzees living in the Gombe Stream Game Reserve in Tanzania, and the families of the animals she originally followed are now entering their third generation. Since she was the first to undertake a long-term study of humankind's closest living relative in the wild, her findings on raising of the young, social relationships, and aggression have had important implications not only for students of primate behavior, but also for anthropologists. For the young reader her remarkable story is one of adventure and plain hard work laced with danger and humor.

HAGAN MEMORIAL LIBRARY
CUMBERLAND COLLEGE
WILLIAMSBURG, KENTUCKY 40769

Jubilee is a soft, fuzzy stuffed toy. She is named after the first chimpanzee born in the London zoo. Mrs. Goodall gave the toy chimp to her baby, Jane, when Jane was one year old. Jubilee is worn and old now, but for more than 40 years Jane Goodall has kept her toy. She has spent much of her life with real chimps in Africa.

Jane was born April 3, 1934, in Bournemouth, England, to Mortimer and Vanne Morris-Goodall. She has wanted to study wild animals ever since she was eight years old. Even then, she knew she would go to Africa one day.

When she was a little girl, Jane brought home all sorts of pets. One of them was just an earthworm. Jane saw something special even in that worm. When she put it on the soft earth, she could see how it moved its body to wriggle into the ground again.

Jane said she had a "natural history museum." In it were living things like tortoises and guinea pigs as well as birds' eggs and sea shells. Jane charged money to see them, but it was not for herself. She had heard that there was a farm that took care of old, unwanted horses. This was where her pennies went.

Jane didn't collect animals, though, the way some people collect stamps. Her goal was to find out how they behaved. Once, when Jane had been missing for five hours, her mother called the

police. Jane didn't know they were looking for her. All that time she was sitting in a hen house nearby. For five hours she had been waiting there in the hope that a hen would lay an egg!

Jane told her mother about her dream of going to Africa. Mrs. Goodall said that she would have to study very hard. To learn all she needed to know to be a scientist, Jane had to go to the university.

It didn't happen. Jane's parents were divorced while she was in high school. There was no money to send Jane to the university. Jane went to London instead, where she worked in an office.

One day Jane got a letter from a friend of hers who lived in Kenya, Africa. She asked Jane to visit her at her family's farm. Jane went back home so that she could save money for her trip. She paid for her plane fare by working as a waitress. In Kenya she stayed with her friend for a month. Then she had to look for work again. She wanted to stay in Kenya. She kept on thinking and talking about animals.

Someone told Jane that she should meet Dr. Louis Leakey. He was head of Kenya's National Museum of Natural History. Dr. Leakey and his wife Mary were famous scientists. They had spent years looking for signs of prehumans, the first humanlike beings on earth. What they found were not the bones of the first people who had lived, but the beings from which humans came.

Dr. Leakey offered Jane a job. She jumped at the chance. First she worked in his office. Then she was asked to go with the Leakeys on their next trip. Jane was thrilled. Not only would she learn more about their work, but she would also have a chance to see animals living in the wild.

She was not disappointed. Tiny antelope bounded over the dusty Serengeti plains. Giraffes grazed on leafy tree tops. Once Jane came face to face with a young lion. She was very frightened. But she did not run. Slowly she climbed the hill toward camp. The lion watched her a long time, then wandered away.

The Leakeys trained Jane to look for even the smallest bone or stone that might have been used by prehumans. It was like a dream. Jane held in her hand the remains of beings, something like herself, who had lived millions of years ago. What did they look like? How did they live? In what ways did they act like those who came after them?

No one could answer these questions. The hints that the humanlike beings left behind were too few. Still, there was a way to find out something about them. Dr. Leakey told Jane about a group of chimps living on the shores of Lake Tanganyika. Their stretch of lakeshore was in Tanzania, the country next to Kenya. Dr. Leakey hoped that someone could study them while the land was still wild. The way they lived might shed some light on the way prehumans had lived long, long ago. Dr. Leakey asked Jane if she wanted to take on the job. Of course she did!

Someone would have to pay for such a study. The Leakeys' money went for their own work. It would be hard to convince others that an untrained young person could do the job. It took weeks of writing and waiting before the right answer came. Jane had gone back to England.

The Wilkie Foundation would help, but there was not much money. The study was planned to last only six months. No matter. It was a chance for Jane to prove that she could do such work.

Now there was just one more problem. The Tanzanian government would not let her go into the wilds alone. Jane's funds couldn't pay for a helper. It all seemed lost. But Mrs. Goodall knew how very much the trip meant to her daughter. As it turned out, she made it possible for Jane to go by going herself. The two women packed their gear, and Jane was on her way at last.

The place where they would go was Gombe Stream, 12 miles away from Kigoma. There, except for a few fishermen and Dominic, their cook, they would be all alone. The only way they would have to get back to the village was in a 12-foot boat.

As they cruised along the shore of the great lake, Jane watched the land change. First there were bare rocky points and deep, narrow bays. Small groves of trees grew in the steep valleys. Every now and then she saw a few mud and grass huts clinging to the rocky slopes. Then, where the chimps lived, the mountains were clothed in trees. The valleys were choked with lush, green growth.

It was wild, fierce country. Mrs. Goodall couldn't see how anyone could climb the steep slopes or struggle through the thick jungle. She was worried. They knew nothing about life in the wild. How would they manage?

She wasn't the only one who was worried. David Anstey, the game ranger, planned to stay a day or two until they got settled. He would see that they met some of the fishermen. He would ask one of them to be Jane's guide and another to be her porter. The porter would carry her gear for her. This was their way. They did not want strangers in their land alone.

Jane argued. She wanted to be alone while observing the animals. As soon as they had set up camp, she slipped away. She climbed up a mountain and stretched out on a big flat rock. There she could look out over the great lake and the green forest below. She saw a troop of baboons. A little bushbuck dashed away at the sight of her. Jane was at peace.

The next day Jane set out with her guide and porter. They headed up a steep valley. Big trees closed off the sky. The path tunneled beside a

stream of water racing from the cliffs above them. Birds flashed by. Monkeys screamed from the branches.

Her guide stopped under a huge tree. It was full of small red fruit. The ground below was littered with branches. This was where the chimps had been feeding. Jane whispered that she wanted to stay and watch, but not too close.

She sat down to wait. She had binoculars to help her see in the distance. Soon she heard the "pant-hoots" of a group of chimps. The hoot was followed by a loud sucking in of breath. She had heard these same calls in London at the zoo but never in such numbers.

Then a chimp came out of the forest and climbed up the tree. Then another. And another. Jane saw a baby clinging to its mother's tummy. She counted 16 chimps climbing up the tree.

That was all she saw. Once the chimps were in the tree, they were hidden by its branches. Once in a while she would see a bit of fruit or a twig drop down. She moved a little closer. They heard her. One of the chimps sounded the alarm. The whole troop swung down from the branches and raced away.

Chimps came every day until they had eaten all the fruit on the tree. Then, for two months, Jane was unable to learn more about them. They fled as soon as she and the two men spotted them. Day after day, Jane came back to camp without seeing a chimp.

At a time when Jane had little hope of success, she and her mother came down with malaria. For days they were too weak from fever to leave their tent. Dominic urged them to go back to the village so that a doctor could take care of them. Jane said no. They were too sick to stand the 3-hour trip in their little boat. Mrs. Goodall's temperature peaked at 105 degrees. Dominic cared for the women as best he could.

The first morning that she was strong enough to leave her bed, Jane crept off by herself. She went to "the Peak," the mountain that she had climbed the day they set up camp. It took a long time to reach the top, and she was very, very tired. She lay down to rest, 1,000 feet above the lake.

Three chimps stopped to look at Jane lying on her rocky bed. This time they did not run. After they had seen their fill, they slowly moved away.

That day was the turning point in her study. The chimps were not so afraid of her when she was by herself and stayed very still.

The chimps came to the fig trees below the Peak almost daily. Jane hauled a tin trunk to her look-out. Inside were a coffee pot, a few cans of beans, and a blanket. She spent the night at the Peak when the chimps were sleeping nearby. It saved her an hour's climb the next day. Her guide and porter did not feel they had to go with her, since they knew where she was.

Jane started giving the chimps names. There was David Graybeard who, of course, wore a short gray beard on his chin. Goliath was big and strong. She knew Flo by her ragged ears. She named Mr. McGregor for his bald head and Ollie for a hairdo that looked something like her Aunt Olwen's.

Chimps in the wild acted like humans in some ways, and they surely hadn't learned those ways from people. They kissed. They held hands. They patted each other when one of them was upset. They tickled and cuddled their babies. "There is something rather funny about it," Jane said later. "One has the feeling that . . . one is looking at one's long-lost cousins."

Already she had found out something that no one had known before. Chimps eat meat. One day Jane came upon three chimps—a male, a female, and a youngster. They were sharing a piglet one of them had caught. Later, Jane would see chimps hunt. It seemed that once in a while, meat was a big treat for the chimps.

Ants were another treat. Once Jane saw David Graybeard come across an ant hill. But he did not stamp on it and wreck it, as one might expect. That would have sent the ants away. Instead, he picked a branch from a tree. Then he carefully stuck the branch into the nest. The ants

clung to the stick. When David pulled the stick out, his dessert was waiting for him. After this, Jane saw other chimps "fishing" for ants. Young chimps copied the grown-ups in their play, just as young humans learn how to use tools.

Jane also saw chimps chew leaves to make them wet. This made a wad of leaves act like a sponge. They used these "sponges" to sop up water from shallow holes.

Sponges and fishing poles. Perhaps this was how humans—or prehumans—had started shaping things around them for their own use. Dr. Leakey would want to hear about this. Jane's notebooks were full of what she had seen.

Untrained though she had been, now Jane's work was seen as important. The National Geographic Society agreed to pay for her studies at Gombe. Dr. Leakey was able to have her accepted by Cambridge University. Once in a great while, Cambridge would let someone work for a doctoral degree who had not gotten a college degree first. Jane was seen as a person who could do this. For the next few years she went back and forth between her studies at Gombe and her studies at Cambridge in England. Thanks to the chimps, she became Dr. Jane Goodall.

Jane was pleased that, the first time she came back from England, the chimps still knew her. They had grown used to this hairless creature who sat so silently watching them. The bravest of the lot even came calling.

It happened one time when Jane was back in the hills. David Graybeard came down to camp. There was a palm tree that shaded Jane's tent. It was bearing oil nuts. David spent an hour eating

right above Jane's home. The next day he did it again.

A short time later Jane was sitting in front of her tent when David came to call. He walked up to her and stopped about five feet away. Then he stood straight up. His hair began to stand on end. He looked huge and very fierce. All at once he rushed toward Jane. He snatched a banana from the table next to her and hurried off to eat it. Bit by bit his hair lay down to make his normal sleek coat. He hooted noisily. The friendship had started.

After that, Jane asked Dominic to put out bananas for the chimps. Soon David's friend Goliath was coming to the camp. Then shy old William joined them.

The National Geographic Society wanted photos of the chimps and hired a young man to take pictures. His name was Hugo van Lawick. Hugo had made a film of Dr. Leakey's work in Kenya. Like Jane, he wanted to work with animals. Dr. Leakey wrote to Mrs. Goodall that he had "found someone just right as a husband for Jane."

Jane was worried that a stranger loaded down with cameras would scare the chimps away. She was wrong. Strangely, it was David Graybeard who helped. He came up to Hugo and asked for a banana. The other chimps watched the strange man and David. He helped them feel safe.

Jane built blinds in the forest for Hugo and his cameras to hide in. If Hugo was hidden, the chimps would act as they do in the wild when humans are not around. Hugo was sure that they could not see him. Then, one day, just as he was starting to take a picture, his camera moved. A big hairy hand was pulling at his shirt! David had found his hiding place.

When Hugo left just before Christmas, Jane missed him very much. Still, she dressed up a little tree with pretty paper and cotton balls. On Christmas morning she put a big pile of bananas around the tree. David and Goliath jumped for joy. They screamed and hugged each other. Jane was very quiet.

The next Christmas it was her turn to leave Hugo behind. She went back to Cambridge. Jane had been gone for a week when Hugo sent her a wire. "Will you marry me?" it said. Jane said yes. A statue of David Graybeard stood on top of their wedding cake.

The camp at Gombe started to grow. Students came to be Jane's helpers and study the chimps.

The chimps themselves started getting in the way. They fought over the food at the camp. Laundry hung on lines to dry was never safe. More than once Jane had to climb trees to find her clothing. But she gave orders that they were

never to be scolded. This might frighten the whole troop away, and her work would be at an end.

Jane came to see that she had made a mistake in touching and playing with the chimps. They

had lost their fear of humans. And a full-grown male chimp is three times as strong as a man.

So that the chimps would not harm the fishermen, they moved their camp away from the shore. Jane made a rule that no one could touch the chimps. They started putting bananas in boxes with covers. Then the boxes became fancy feeding stations. As soon as the chimps learned how to get the bananas out, Jane had to figure out a new way to keep the bananas in. Later they had a building and could open and close the stations from indoors.

An angry chimp was not the only danger at Gombe. One day Ollie's new baby was sick. Soon it died. Ollie carried her dead baby around for a day or so. She seemed dazed and sad. Then Jane found out that people nearby had come down with polio. She knew that a disease could be passed from humans to chimps. Jane was afraid. She herself was going to have a baby.

Polio vaccine was rushed to Gombe. The people at the camp were given shots. The chimps were given the vaccine in food. It may have kept some of them from getting sick.

It was a terrible time. A chimp would not be around for a few days. Then it might creep back to Gombe with its arms or legs useless. Two of the chimps had to be killed because they could not live on their own. Others died by themselves in the forest.

Jane's baby was born in 1967. He was named Hugo after his father, but his parents call him "Grub." This means "bush baby" in Swahili, the language of the people who live near Gombe.

Jane learned her lessons in mothering from the chimps. She seldom left Grub alone when he was small. She carried him in a sling on her back.

"I cuddled Grub lots," Jane says. "Chimps keep their babies happy by cuddling them whenever they want it." She tried not to scold him or stop him from doing something naughty. Instead, she would turn his attention to something else.

To Jane, Gombe was a good place to raise a child. In her words, "You can focus on what's important instead of spilled food and other things that don't matter."

Still, there were dangers for a human child. The first sentence Grub learned was, "Lion out there eat me." There was a chance that a chimp might choose Grub for dinner, too. It happened rarely, but chimps had been known to steal human babies.

To keep him safe, his parents built a big, sturdy cage by the lake. The chimps did not come to that stretch of beach. There he could play by himself. At other times his parents stayed with him. They taught Grub to respect wild animals.

Grub felt safe in the forest, but he knew how to be careful. When he was 3 years old, he visited

Mrs. Goodall in England. When he went into her garden, he asked her, "Is it safe for me to be here?" Then his ball rolled under a bush. He hit the plant with a big stick to make sure there were no snakes lurking under its branches. Only then did he go after the ball.

Jane, Hugo, and Grub traveled to other places in Africa. Often, Hugo was gone for a long time when he was sent somewhere to take wild life pictures. Jane hated to stay away from her own work. As time passed, she and Hugo were apart more and more. They decided that a divorce would be best for them.

After the divorce they still worked together. "We're even better friends now," Jane says. "We don't feel any bitterness. We had a jolly good time and produced Grub. We achieved together more than we could have done separately."

In 1972 Jane joined a group to ask the government of Tanzania to make Gombe a national park. She was not looking forward to the trip. She had heard that the Tanzanian park director was a hard man to get along with.

His name was known to many. Derek Bryceson had been a member of the Royal Air Force in World War II. His plane had been shot down. He was badly hurt. Doctors had told him he would never walk again, but, after many

months he was able to get around with a cane.

Bryceson had come to Africa to farm. He helped his adopted country gain its independence. Now he was serving in Tanzania's National Assembly.

Jane found that like her, Bryceson wanted to protect wildlife. He had been trying to stop farmers from taking over more land. They must learn to make better use of the land they had, he said. Instead of being against Jane's idea to make Gombe into a park, he was very much for it. Thanks in part to his efforts, Gombe did become a national park.

Jane married Derek Bryceson two years later, but their time together was all too short. He died of cancer in the fall of 1980. Jane's sadness was echoed by the whole country. Her husband had been one of Tanzania's best-loved lawmakers.

During these years, life at Gombe showed its darker side. In 1975, a group of rebels from Zaire crossed the lake and attacked the camp. They captured four students who were helping Jane and held them until a ransom was paid.

Jane also found rebels among the chimps. A female named Passion attacked a mother chimp and her child. She killed the baby and ate it. Other babies had disappeared. No one knew what had happened. Now the researchers thought Passion might have killed them.

They kept a close watch on her. When she tried to get too close to other babies, she was frightened away. The next year Passion had a baby of her own, and the killings stopped.

Then Jane found out something even stranger. A band of ten animals broke away from the main group. They moved to the south. At first the

two bands got along with each other. Males from each band patrolled the "border." They scared away any chimp from the other band who tried to cross it. Later, chimps from the Gombe band ganged up on a chimp from the southern band. They killed him. After three years all the males in the southern band had been killed or were missing. The females went back to the northern group.

This was the first time anyone had watched a

war between two bands of animals. It lasted for more than three years.

Jane's work has always been this way. Just when she began to feel that she knew all there was to know about the Gombe chimps, they would surprise her. Now she has completed 20 years of study. There is still much to be learned.

Now Jane is watching the grandchildren of the first chimps she met. David Graybeard, Goliath, Flo, and Ollie have all died. After David, Mike took over the group. He was a small chimp, Jane remembers, but he was a great bluffer. He used to bang on empty oil cans and make an awful noise. He frightened everybody into thinking he was the strongest. After Mike, Flo's son Figan became top chimp.

And Jane is tops in her field. She has written books and taught at famous universities. She has been featured in nature films. She has given speeches all over the world.

Jane has learned important lessons from the animals she has so patiently watched. And she

passes on what she has learned to others. She is one of those rare people who love their work and share it. "I don't think there's a day when I don't wake up and think how lucky I am," she says.

The Author

Mary Virginia Fox is the author of many books and articles for adults as well as young people, including several books in the Taking Part series. During her long career, she has lived in a number of foreign countries, including Colombia, Tunisia, Iran, and the Philippines, where she worked as a newspaper columnist. Ms. Fox is a member of the American Society of Journalists and Authors and the National League of American Pen Women. An honors graduate of Northwestern University, she now lives with her husband in Middleton, Wisconsin.

The Illustrator

Nona Hengen has illustrated magazine articles about animals and several books. Now a free-lance artist, she formerly taught visual communications and instructional systems technology at the university level and worked as a director of educational media services.

She was graduated from Fort Wright College in Spokane, Washington, and received a Ph.D. in education from Indiana University.

Cover photograph by Baron Hugo van Lawick, ©National Geographic Society.

HAGAN MEMORIAL LIBRARY
CUMBERLAND COLLEGE
WILLIAMSBURG, KENTUCKY 40769